A++ and the Lambda Calculus

Georg P. Loczewski

A++ and the Lambda Calculus

Principles of Functional Programming

www.tredition.de

IMPRESSUM

Copyright ©2018 Georg P. Loczewski
A++ and the Lambda Calculus

The book was set by the author using the LATEX typesetting system and was printed and bound in the Federal Republic of Germany.

1st. Edition 2018
tredition GmbH, Hamburg

ISBN
978-3-7469-3811-0 (Paperback)
978-3-7469-3809-7 (Hardcover)
978-3-7469-3810-3 (e-Book)

See also *A++ The Smallest Programming Language in the World*[29]

Printed in Germany

   The author and publisher make no warranty of any kind, expressed or implied, with regard to these programs or the documentation contained in this book. The author and publisher shall not be liable in any event for incidental or consequential damages in connection with the use of these programs.

   All rights reserved. No part of this book may be reproduced in any form by any electronic or mechanical means (including photocopying, recording, or information storage and retrieval) without permission in writing from the publisher and the author.

   The book was set by the author using the LATEX typesetting system and was printed and bound in the Federal Republic of Germany.

To my wife Ursula
and my sons Thomas and Johannes
dedicated in love.

Contents

I The Lambda Calculus   1

1 Introduction   3
    1.1 Origin 3
    1.2 Definition 3
    1.3 Literature 3
    1.4 Syntax of Lambda Expressions 4

2 Basic Rules for Lambda Conversions   5
    2.1 Notation used in Conversion Rules 5
       2.1.1 Notation used to specify conversion of lambda expressions 5
       2.1.2 Notation used to specifiy substitution 6
    2.2 Alpha Conversion 6
       2.2.1 Rules of Alpha Conversion 6
       2.2.2 Beta Conversion 7
       2.2.3 Rule of β-Conversion 7
       2.2.4 β-Reduction 7
       2.2.5 β-Abstraction 8
    2.3 Eta Conversion 8
       2.3.1 η-Reduction 8
       2.3.2 η-Abstraction 9

2.4		Rules of Associativity	9
	2.4.1	Rule of Associativity for Abstraction	9
	2.4.2	Rule of Associativity for Application	10
	2.4.3	Example for both rules:	10
2.5		Y-Combinator	10
	2.5.1	Using the Y-Combinator to implement recursion	12

II A++   15

1 Introduction to A++   17

- 1.1 Purpose of A++ and Origin 17
  - 1.1.1 Purpose 17
  - 1.1.2 Origin 17
  - 1.1.3 ARS — *Generalization of the Lambda-Calculus* 18
  - 1.1.4 Name of the language 19
- 1.2 Motivations for the development of A++ 19
  - 1.2.1 To support an alternate method of teaching the principles of programming 19
  - 1.2.2 To provide a learning tool for exploring the fundamentals of programming 19
  - 1.2.3 To support a method teaching powerful programming patterns applicable to most languages 20
  - 1.2.4 To open a new view of programming for many programmers: 20
- 1.3 Features of A++ 20
  - 1.3.1 Programming paradigms supported: 20
  - 1.3.2 Constitutive Principles of A++ 20

		1.3.3	Closure	22
		1.3.4	Basic abstractions derived from ARS	22
		1.3.5	Development of Applications with A++	23
	1.4	Internal Architecture of A++		24
		1.4.1	Overview	24
		1.4.2	Internal Structure Overview (commented version)	25
		1.4.3	Syntax of ARS (A++)	26
		1.4.4	Definition of A++ in the form of a Tree Diagram	27
		1.4.5	Commenting the definition:	27
		1.4.6	Examples of A++ - Syntax	28

2 General Programming Patterns and A++ — 31

	2.1	Closure Pattern	31
	2.2	CLAM Pattern	32
	2.3	List Pattern	32
	2.4	Dictionary Pattern	34
	2.5	Set Pattern	35
	2.6	Recursion Pattern	35
	2.7	Higher Order Function Pattern	36
	2.8	Message Passing Pattern	36
		2.8.1 Classes of objects	36
		2.8.2 Instance of a class	38
		2.8.3 Constructors	38
		2.8.4 Creating instances of a class	39
		2.8.5 Sending messages	40
		2.8.6 Executing methods	41

	2.8.7	Essential features of object oriented programming	43
	2.8.8	Relation between classes	44
2.9	Meta Object Protocol Pattern	44

Bibliography 45

Part I

The Lambda Calculus

Chapter 1

Introduction

1.1 Origin

The **Lambda Calculus** has been created by the American logician **Alonzo Church** in the 1930's and is documented in his works published in 1941 under the title *'The Calculi of Lambda Conversion'*.

Alonzo Church wanted to formulate a mathematical logical system and had no intent to create a programming language. The intrinsic relationship of his system to programming was discovered much later in a time in which programming of computers became an issue.

1.2 Definition

> DEFINITION 1 (LAMBDA CALCULUS)
> The Lambda Calculus defines the laws for the formulation and conversion of lambda expressions.

1.3 Literature

As a *mathematical logical system* the Lambda Calculus is covered in detail in [2] and less comprehensively but in a more readable form in [34]. A clear account of the historical origins and basic properties of the lambda calculus is presented by Curry and Fey in their book [12]. This view is taken from [21] page 23.

From the programmer's point of view the Lambda Calculus is adressed in [21], [22], [3].

1.4 Syntax of Lambda Expressions

The syntax of lambda expressions is defined as follows:

> DEFINITION 2 (SYNTAX OF A LAMBDA EXPRESSION)
> t is a lambda expression, if
>
> - $t = x$ where $x \in Var$,
>   or
> - $t = \lambda x.M$ where $x \in Var$ and M is a lambda expression,
>   or
> - $t = (MN)$ where M and N are lambda expressions.

The Lambda Calculus therefore includes three diffenrent types of lambda expressions:

- **variables** (referencing lambda expressions)
- **lambda abstractions** (defining functions)
- **applications** (invoking functions)

Remark:

The parentheses in the syntax of an application are not mandatory. This results from the law of associativity for applications introduced below.

Chapter 2

Basic Rules for Lambda Conversions

2.1 Notation used in Conversion Rules

The notation used to specify a conversion of a lambda expression and associated with it substitutions to be performed varies from textbook to textbook. In 'Programmierung pur' we adopted the notation used by Chazarain in [3]. For greater clarity and flexibility we prefer to use the following notation here:

2.1.1 Notation used to specify conversion of lambda expressions

---
NOTATION 1 (CONVERSION OPERATOR)

$M \underset{\alpha}{\rightarrow} N$ rule applies to conversion from left to right

$M \underset{\alpha}{\leftarrow} N$ rule applies to conversion from right to left

$M \underset{\alpha}{\leftrightarrow} N$ rule applies to conversion in both directions

where the Greek letter below the arrow specifies the type of conversion.

2.1.2 Notation used to specifiy substitution

> NOTATION 2 (SUBSTITUTION OPERATOR)
>
> $M[x \rightarrow N]$ x in M is substituted by N
>
> $M[x \leftarrow N]$ N in M is substituted by x
>
> $M[x \leftrightarrow N]$ x in M is substituted by N
> or N in M is substituted by x
> depending on the direction of conversion

2.2 Alpha Conversion

Bound and Free Variables

> DEFINITION 3 (BOUND AND FREE VARIABLES)
> In the following lambda expression:
>
> $$\lambda x.xy$$
>
> "y" is a free variable whereas "x" is a bound variable. Free variables take their values from lambda expressions on higher levels.

2.2.1 Rules of Alpha Conversion

An Alpha Conversion of a lambda expression is defined as follows:

> RULES 1-3 (ALPHA CONVERSION)
>
> $$\lambda x.M \underset{\alpha}{\rightarrow} \lambda x_0.M[x \rightarrow x_0] \quad\quad 1$$
>
> $$\lambda x.M \underset{\alpha}{\leftarrow} \lambda x_0.M[x \leftarrow x_0] \quad\quad 2$$
>
> $$\lambda x.M \underset{\alpha}{\leftrightarrow} \lambda x_0.M[x \leftrightarrow x_0] \quad\quad 3$$
>
> where x_0 is not allowed to be a free variable in M.
> The process of alpha conversion may not alter the value of the expression. The expression to be converted (M) and the converted result (N) are said to be equal *modulo alpha*:
>
> $$M =_\alpha N.$$

2.2. ALPHA CONVERSION

The third rule for alpha conversion is a combination of the first two rules.

Alpha conversion may be required, if substitutions are performed in lambda expressions.

2.2.2 Beta Conversion

β-Conversion primarily consists of the process of substituting a bound variable in the body of a lambda abstraction by the argument passed to the function whenever it is applied. This process is called β-*reduction*.

The inverse process to convert a β-reduced lambda expression back to the reducible expression is another aspect of β-conversion and is called β-*abstraction*.

2.2.3 Rule of β-Conversion

> RULE 4 (BETA CONVERSION)
> The following transformation is called β-conversion:
> $$((\lambda x.M)N) \underset{\beta}{\leftrightarrow} M[x \leftrightarrow N]$$

2.2.4 β-Reduction

Reducible Expression 'redex'

β-reduction can be applied only to reducible expressions. A reducible expression called 'redex' for short is defined as follows:

> DEFINITION 4 (REDEX)
> A redex is a reducible expression and is represented as:
> $$((\lambda x.M)N)$$

Rule of β-Reduction

> RULE 5 (BETA REDUCTION)
> The following transformation is called β-reduction:
> $$((\lambda x.M)N) \underset{\beta}{\rightarrow} M[x \rightarrow N]$$

2.2.5 β-Abstraction

Rule of β-Abstraction

> RULE 6 (BETA ABSTRACTION)
> The following transformation is called β-abstraction:
> $$((\lambda x.M)N) \underset{\beta}{\leftarrow} M[x \leftarrow N]$$

Examples:

- $((\lambda x.x\ x)(\lambda y.y)) \rightarrow_\beta ((\lambda y.y)(\lambda y.y)) \rightarrow_\beta (\lambda y.y)$
- $((\lambda x.(\lambda y.x\ y))y) \rightarrow_\beta (\lambda y_0.y\ y_0)$

Remark: The second example demonstrates the necessity of alpha conversion. The lambda bound variable y had to be renamed y_0, to prevent capturing of the free y (resulting from the substitution of x by y in the body of the first lambda absstraction) by the second lambda.

2.3 Eta Conversion

Like β-conversion η-conversion can be performed from left to right and from right to left and is therefore subdivided in

- η-reduction and
- η-abstraction

> RULE 7 (ETA CONVERSION)
> The following transformation of a lambda expression is called η-conversion.
> $$(\lambda x.Mx) \underset{\eta}{\leftrightarrow} M\ ,$$
> where x may not be a free variable in M.

2.3.1 η-Reduction

η-reduction is useful to eliminate redundant lambda abstractions. The following rule can be interpreted like this:

If the sole purpose of a lambda abstraction is to pass its argument to another function, then the lambda abstraction is redundant and can be stripped via η-reduction.

In an environment where *'eager evaluation'* is used like in Scheme such redundant lambda abstractions are used as a wrapper around a lambda expression to prevent immediate evaluation.

> RULE 8 (ETA REDUCTION)
> The following transformation of a lambda expression is called η-reduction.
> $$(\lambda x.Mx) \underset{\eta}{\to} M ,$$
> where x may not be a free variable in M.

2.3.2 η-Abstraction

η-abstraction on the contrary is useful in 'eager' languages to create a wrapper around a lambda-expression. In 'lazy' languages like Lambda Calculus, A++, SML, Haskell, Miranda etc. η-conversion, abstraction and reduction alike, are mainly used within compilers. (See [21] page 22.)

> RULE 9 (ETA ABSTRACTION)
> The following transformation of a lambda expression is called η-abstraction.
> $$(\lambda x.Mx) \underset{\eta}{\leftarrow} M ,$$
> where x may not be a free variable in M.

2.4 Rules of Associativity

2.4.1 Rule of Associativity for Abstraction

> RULE 10 (RULE OF ASSOCIATIVITY FOR ABSTRACTION)
> Abstraction is associative from left to the right.

Example:

The expression

$$\lambda x.\lambda y.\lambda z.M$$

can be rewritten as:

$$\lambda xyz.M$$

2.4.2 Rule of Associativity for Application

> RULE 11 (RULE OF ASSOCIATIVITY FOR APPLICATION)
> Application (Synthesis) is associative from right to left.

Example:

The expression:

$$((MN)P)$$

can be rewritten as:

$$MNP.$$

2.4.3 Example for both rules:

$$\lambda x.\lambda y.((xy)z)$$

is equivalent to

$$\lambda xy.xyz.$$

2.5 Y-Combinator

Definition

The "Y-combinator", which is sometimes also called *fixpoint combinator*, was discovered by *H. Curry*. With its help it is possible to *handle recursive functions* in the Lambda Calculus.

2.5. Y-COMBINATOR

In many programming languages it is possible to simply refer to the name of a function within the function itself. This is called *implicit recursion*, which is not possible in the Lambda Calculus because all lambda abstractions are by definition 'anonymous functions'.

> DEFINITION 5 (Y-COMBINATOR)
> The Y-Combinator is defined as follows:
> $$Y = \lambda f.((\lambda x.(f(x\ x)))(\lambda x.(f(x\ x))))$$

Basic Usage of Y-Combinator

Fixpoint of a Function

If it can be shown that $(M\ N) =_\beta N$ is true, then N is called either a *'fixed point'* or a *'fixpoint'* of M.

According to H. Curry, there exists a function, that generates such a *fixed point* of M. This function is the so-called Y-combinator introduced above.

> DEFINITION 6 (FIXPOINT OF A FUNCTION)
> $$(MN) \underset{\beta}{\rightarrow} N$$

According to the above definition a fixpoint of a function can be seen as a lambda expression that, if passed as argument to this function is returned by the function.

Generation of fixpoint using Y-combinator By definition the Y-combinator has the function to generate a fixpoint of M:

> RULE 12 (FIXPOINT GENERATION USING Y-COMBINATOR)
> $$(M(Y\ M)) \underset{\beta}{\rightarrow} (Y\ M)$$

Fixpoint expansion by β-reduction It looks like magic that expansion can be obtained by reduction as well as by abstraction. The first of these alternatives is shown below:

> RULE 13 (FIXPOINT EXPANSION)
>
> $(Y\ M) \underset{\beta}{\to} (M(Y\ M))$

Verification of the expansion formula:

> PROOF 1 (FIXPOINT EXPANSION)
>
> $(Y\ M) \to_\beta ((\lambda x.(M(x\ x)))(\lambda x.(M(x\ x))))$
> $\quad \to_\beta (M((\lambda x.(M(x\ x)))(\lambda x.(M(x\ x)))))$
> $\quad \to_\beta (M(Y\ M)).$

2.5.1 Using the Y-Combinator to implement recursion

Introduction

The following classical example for recursive programming, the function to calculate the factorial of a natural number, is used to test the Y-combinator.

In order to simplify the procedure we use the functions $IF, =, *, -$ as predefined 'lambda abstractions' and assume as well the availability of all natural numbers as values. To make these definitions in the Lambda Calculus would not be difficult, but it would certainly distract from our present topic.

Steps to Eliminate Implicit Recursion

1. recursive function to calculate the factorial including illegal implicit recursion:
   $FAC = \lambda n.(IF(= n\ 0)1(*n(FAC(-n\ 1))))$

2. locating illegal implicit recursion:
   $FAC = \lambda n.(...FAC...)$

3. eliminating illegal implicit recursion:
   $FAC = (\lambda fac.\lambda n.(\ldots fac \ldots)FAC)$

4. simplified representation:
   $FAC = (M\ FAC)$
   where:

2.5. Y-COMBINATOR

$M = \lambda fac.\lambda n.(...fac...)$

The recursion is eliminated by enclosing the recursive function in a lambda abstraction, that receives the function to be invoked recursively as an argument.

5. Using definition 12 and the result in step 4 FAC is identified as a *fixed point* of M.

    Therefore we can write:

    $FAC = (Y\ M)$

6. The function to calculate the factorial can now be rewritten as:

    $FAC = (Y\ \lambda fac.\lambda n.(...fac...))$

7. To compute the factorial of n the following expression has to be evaluated:

    $((Y\ \lambda fac.\lambda n.(...fac...)\ )n)$

Example of Step by Step Evaluation

- predefined variables:
    - $FAC = (YM)$
    - $M = \lambda fac.\lambda n.(IF(= n\ 0)1(*n(fac(-n\ 1))))$
- expression to be evaluated:

    $(FAC\ 1)$

- expression rewritten replacing FAC, a fixed point of M, by (YM):

    $((YM)1)$

- expression rewritten using rule 12:

    $((M(YM))1)$

- expression rewritten substituting the 1st occurrence of M::

    $((\lambda fac.\lambda n.(IF(= n\ 0)1(*n(fac(-n\ 1)))))(YM))1)$

- performing $_\beta$-reduction by substituting variable fac in the body of the lambda abstraction by the argument $(Y\ M)$:

    $(\lambda n.(IF(= n\ 0)1(*n((YM)(-n\ 1))))1)$

- performing β-reduction by substituting variable n in the body of the lambda abstraction by the argument 1:

  $(IF(=1\ 0)1(*1((YM)(-1\ 1))))$

- evaluating the 'IF'-expression by selecting the 'no'-branch and evaluating the subtraction:

  $(*1((YM)0))$

- expression rewritten substituting (YM) by $(M(YM))$ using rule 12:

  $(*1((M(YM))0))$

- expression rewritten substituting the first occurrence of the variable M by its value:

  $(*1((\lambda fac.\lambda n.(IF(=n\ 0)1(*n(fac(-n\ 1))))(YM))0))$

- performing β-reduction substituting 'fac' in the body of the lambda abstraction by the argument (YM):

  $(*1(\lambda n.(IF(=n\ 0)1(*n((YM)(-n\ 1))))0)$

- performing β-reduction substituting n in the body of the lambda abstraction by the argument 0:

  $(*1((IF(=0\ 0)1(*0((YM)(-0\ 1))))))$

- evaluating 'IF'-expression returning the 'yes'-branch:

  $(*1\ 1)$

- evaluating the multiplication:

  1

Part II

A++

Chapter 1

Introduction to A++

1.1 Purpose of A++ and Origin

1.1.1 Purpose

A++ is a *minimal programming language* that has been built on the **Lambda Calculus** with the purpose to serve as a learning instrument rather than as a programming language used to solve practical problems.

A++ is introduced as a **universal learning tool for programming**, confronting students with the essence of programming and helping to master this confrontation.

It is also supposed to help become thoroughly familiar with programming patterns that can be applied in other languages needed to face the real world. Learning of new programming languages will be a lot easier and need less time after an intensive training in A++, leading to earlier productivity in the new programming language.

1.1.2 Origin

A++ is a language that owes it's existence to a *generalization of the base operations of the Lambda Calculus* and may very well be called the *smallest programming language in the world*.

A++ has been developed in 2002 in the context of 'ARS based programming' covered in detail in the book 'Programmierung pur' (ISBN 3-87820-108-7) with the purpose to serve as a *learning instrument* rather than as a programming language used to solve practical problems.

1.1.3 ARS — *Generalization of the Lambda-Calculus*

The pure **Lambda Calculus** is applicable only to functional programming. **A++** however is built on **ARS** which stands for the basic operations of the Lambda-Calculus in a generalized form. Guy L. Steele, one of the fathers of the Scheme Programming Language, praises the beauty of ARS in his foreword to [33] on page XV and XVI. The following phrase puts everything to a point:

> *Abstraction is all there is to talk about: it is both the object and the means of discussion.* **Guy L. Steele Jr.**

ARS provides a base for imperative programming and object-oriented programming as well and can be applied to programming in almost any programming language.

ARS, the basic operations of the Lambda Calculus *in their generalized form* are defined as follows:

> **Abstraction:** Give something a name.
>
> **Reference:** Reference an abstraction by name.
>
> **Synthesis:** Combine two or more abstractions to create something new.

These operations may sound rather trivial and abstract but taken as principles of programming they change the style and method of programming thoroughly.

The **generalization of the Lambda Calculus** consists in defining the concept of abstraction simply by 'give something a name'. The name hides all the details of the defined. Abstraction thus defined requires an *explicit definition of a name*.

The Lambda Calculus does not allow for an explicit definition of a name. The only possibility to associate a name to a value in the Lambda Calculus is by calling

1.2. MOTIVATIONS FOR THE DEVELOPMENT OF A++

a function with an argument. This operation corresponds to the synthesis operation however and not to the creation of an abstraction. Lambda-abstractions in the Lambda Calculus are 'per se' anonymous.

1.1.4 Name of the language

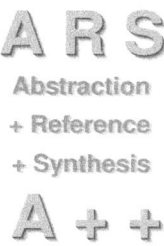

The name **A++** stands for *abstraction plus reference plus synthesis* .
The three elements of the name designate the primitive operations of the language used to build everything else.

1.2 Motivations for the development of A++

1.2.1 To support an alternate method of teaching the principles of programming

A++ supports a method to teach the principles of programming, that leads to a *quick and deep comprehension* of the essence of programming helping as a consequence to considerably *simplify and accelarate* the learning of popular programming languages.

1.2.2 To provide a learning tool for exploring the fundamentals of programming

A++ can be considered a tool providing the possibility to get **hands-on experience** in the process of *learning and exploring the fundamentals of programming*.

1.2.3 To support a method teaching powerful programming patterns applicable to most languages

Learning A++ not only leads an understanding of the essence of programming but also makes students in an early stage familiar with powerful programming patterns that can be applied in almost any language.

The familiarity with these programming patterns learnt with the help of A++ helps substantially to pick up new languages and use them efficiently.

1.2.4 To open a new view of programming for many programmers:

Even **skilled programmers** may considerably benefit from experimenting with A++, because they might never have been exposed before to a *view of programming* which is traditionally at home in artificial intelligence programming and in functional programming languages like Haskell, Miranda, ML, SML, Scheme and Lisp *and not* in commercial programming and languages like FORTRAN, BASIC, COBOL, Pascal, C, C++, Java, etc..

Functional programming languages are considered as programming languages with a *unique expressiveness* that leads to a code, that is more *robust and easier to verify* than in imperative languages.

1.3 Features of A++

1.3.1 Programming paradigms supported:

- **functional programming**,
  (writing expressions to be evaluated),

- **object oriented programming**
  (sending messages to objects),

- **imperative programming**
  (writing statements to be executed), including *structured programming*.

- **logic programming**
  (rule based programming) (ARS++ recommended!)

1.3.2 Constitutive Principles of A++

ARS

- *Abstraction*
  to give something a name,

1.3. FEATURES OF A++

- *+ Reference*
  to reference an abstraction by name,

- *+ Synthesis*
  to combine one abstraction with other abstractions to create something new.

The constitutive principles of A++ are those, that make A++ to what it is. These principles are essentially the nucleus of the language, everything else can be derived from them. ARS, as introduced above in 1.1.3 on page 18 provides three of these principles and 'lexical scope' is the fourth.

A constitutive principle is different from a simple characteristic of a programming language. In this sense ARS is not constitutive in languages like for example Pascal, C and C++. Every programming language must somehow provide a 'name giving' mechanism, a feature allowing to call procedures or functions and the possibility to refer to variables.

In A++ ARS is *universal*, the principles can be applied anywhere at any time because they make up the language.

In most other languages the operations symbolized by ARS can be applied only under certain conditions, only in certain constructs controlled by a complex set of rules, which on top of blocking ARS is different from language to language.

This complex set of rules imposing many restrictions and limitations on the programming activity makes learning how to program more difficult than necessary, distracts students from the essentials of programming, delays the learning process and may prevent students from reaching a profound knowledge of the art of programming combined with a high degree of programming efficiency.

Lexical Scope

'Lexical Scope' defines the access to variables within functions. Variables within functions are either **'lambda-bound'** or **'free'**. The **'lambda-bound'** variables refer to *arguments* passed to the function.

The so-called **'free'** variables must have been defined in the *inherited environment* of the function. In a language with 'Lexical Scope' a lambda abstraction inherits all variables from those abstractions in which it is defined.

> The values of these variables reflect the *time of definition of the lambda abstraction* and not the time of invoking the function!
> The values of the free variables may be changed however and keep their new values indefinitely!

'Lexical Scope' in A++ is coupled with **indefinite extent** in contrast to Pascal where 'Lexical Scope' is coupled with *'limited extent'*.

22   CHAPTER 1. INTRODUCTION TO A++

'Lexical Scope' regulates the access to variables in a lambda-abstraction by its context visible in the program text.

1.3.3 Closure

The 'Closure' principle is so important in A++, because all lambda abstractions are closures, that we would like to list it together with the constitutive principles above. But because all features of a closure can be derived from the principles above, we have to list it separately.

The term closure is an **encapsulation of a lambda-abstraction with its total environment** at the time of creation of the lambda-abstraction. This environment consists of **all the names** that this lambda abstraction has access to.

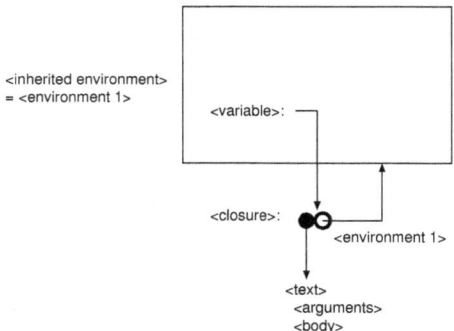

The diagram above uses a technique introduced in [1] representing a closure by two circles. The first one (in our case the black bullet) sympolizes the function and the second one the environment. In our diagram the expressions in the function body have access to all variable bound to the 'lambda', i.e. the arguments and to all variables contained in 'environment 1'.

For more details readers are referred to the discussion of general program patterns in section 2.1 on page 31.

1.3.4 Basic abstractions derived from ARS

The following list is a summary of the most common basic abstractions derived from ARS. A detailed description of each one of them can be found in section ?? on page ??.

- **features directly derived from ARS:**

1.3. FEATURES OF A++

- **logical abstractions**
  (true, false, if, not, and, or),
- **numerical abstractions**
  (natural numbers, *zerop, succ, pred, add, sub, mult)*,
- **relational abstractions**,
  (equaln, gtp, ltp, gep)
- **recursion**,
- creation and processing of **lists**
  (cons, car, cdr, nil, nullp, length, remove, nth, assoc),
- **higher order functions**
  (compose, curry, map, mapc, map2, filter, locate, for-each) ,
- **set operations**
  (memberp, union, addelt),
- **iterative control structure** *('while')*,

- **development of applications** like *'simple account handling'* and *'library management'*.

1.3.5 Development of Applications with A++

The purpose of A++ is not to be used as a programming language to write applications for the needs of the real world. Nevertheless it is possible to write **simple application programs** in A++ like object oriented implementations of a *simple account handling* and a *library management system* .

To write **real world application programs** the language **ARS++** is provided, which extends A++ to a language similiar to Scheme. ARS++ is derived from *ARS plus Scheme plus Extensions* .

The book *Programmierung pur* , at present only available in German under the ISBN 3-87820-108-7, presents **A++** , its usage and implementation in detail and shows as well how programming in languages like ARS++, Scheme, Java, Python, C and C++ can be done in the *spirit of A++* .

The next edition of the book will include **Perl** as well, a powerful programming language in which *ARS Based Programming* can be applied as easily as in Scheme.

The book mentioned above not only describes the language but also presents the interpreters of A++ in 5 programming languages. The A++ - Interpreter in C can be downloaded from the download page of this site.

1.4 Internal Architecture of A++

In pure A++ data and operations on the data are expressed solely by lambda abstractions.

In addition to the elements of the Lambda Calculus symbols (names) are allowed in order to be able to give "something" (i.e. the abstractions) a name.

In spite of the simplicity of pure A++ it is possible to define natural numbers, new syntax constructs (like 'if' and 'while') and even a data construct like 'list' just by building lambda abstractions.

On top of that of course numerous operations can be defined capable of processeing these data (e.g. map,filter,assoc,foreach,car,cons,cdr,sum,add,subtract,multiply and many more).

1.4.1 Overview

The following list represents a coarse uncommented overview of the internal structure of the language. A commented version follows next.

abstraction

- give something a name
    - define
    - name
    - expression
- lambda abstraction
    - lambda
    - list of names
    - list of expressions

reference

- variable name

synthesis

- expression
- list of expressions or void

1.4. INTERNAL ARCHITECTURE OF A++

1.4.2 Internal Structure Overview (commented version)

abstraction

- give something a name
  - define

    Set key in dictionary, i.e. in the lexical environment.
  - name
  - expression

    If name appears in expression eager evaluation will be applied, otherwise lazy evaluation. More comments follow below under 'predefined lambda abstraction'.
- lambda abstraction
  - lambda

    Identifiying keyword.
  - list of names

    List of formal parameters.
  - list of expressions

    Expressions of any kind.

reference

- variable name

  Referring to a name-value pair in the lexical environment.

synthesis

- expression

- list of expressions or void

  List of the actual arguments that will replace the formal parameters of the lambda abstraction. The number of expected arguments depends on the lambda abstraction to be evaluated or the primitive function to be executed and therefore may be 0.

1.4.3 Syntax of ARS (A++)

expression

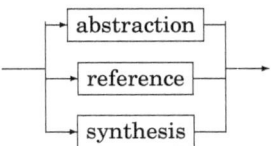

abstraction

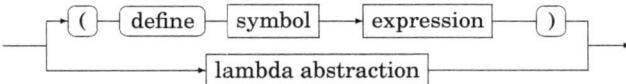

lambda abstraction

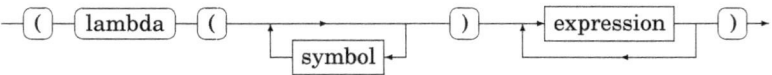

reference

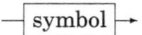

synthesis

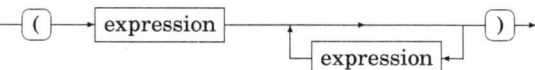

1.4. INTERNAL ARCHITECTURE OF A++

symbol

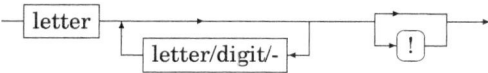

1.4.4 Definition of A++ in the form of a Tree Diagram

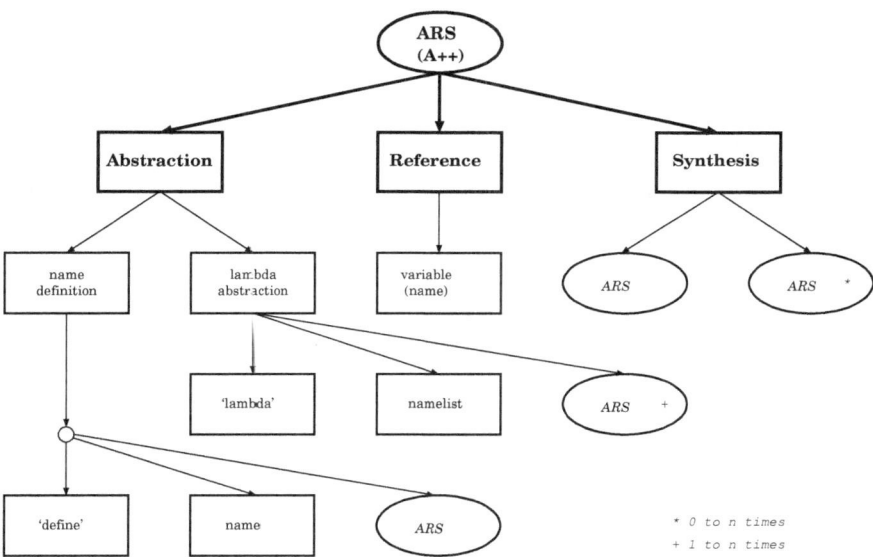

1.4.5 Commenting the definition:

- Like the introduction to the Lambda Calculus the definition of A++ deals with **'lambda expressions'** and defines three different types: 'abstractions', 'references' and 'syntheses'. This corresponds to the 'lambda abstractions', the 'variables' and the 'applications' in the Lambda Calculus. The **few differences** in the A++ - approach are the following:

    - The **syntax of A++** is different. It is borrowed from Scheme and so simple that it can be described in a few words:

> A lambda expression may either be a *symbol* or a construct with *parentheses*. A construct with parentheses represents a lambda abstraction if the first thing following the openening parentheses is the symbol *'lambda'* or *'define'*, otherwise it must be an application.

This is all there is to the syntax! *It can't be simpler.* Even the syntax of the Lambda Calculus is a little bit more complex! (e.g. sometimes parentheses are required, sometimes they are optional.)

- **Abstractions** may be given a name explicitly, matching the general human understanding of 'abstraction' as 'to give something a name'.
- **Abstractions** may contain more than one lambda expression in the body to be evaluated.
- **Applications** may contain more than two lambda abstractions including several arguments passed to the operator.

• The **rules for the conversion** of lambda expressions defined in the Lambda Calculus are valid in A++ as well. Due to *'lazy evaluation'* in A++ lambda expressions can be treated the same way as in the Lambda Calculus.

1.4.6 Examples of A++ - Syntax

Examples of syntax of abstraction 1st alternative in 2.2

```
1 (define first
    (lambda (x y)
3     x))
  ;
5 (define alias name)
  ;
7 (define choice (first a b))
```

Examples of syntax of abstraction 2nd alternative in 2.2

```
1 ;
  (lambda (x)
3   (mult x ten))
  ;
```

Examples of syntax of reference 2.3

```
  first
2 ;
  choice
```

1.4. INTERNAL ARCHITECTURE OF A++

Examples of syntax of synthesis 2.4

```
1 (b t f)
  ;
3 (first a b)
  ;
5 ((lambda(x) (add x three)) two)
```

Chapter 2

General Programming Patterns and A++

A consequent application of *abstraction, reference and synthesis* (**ARS**), leads to programming patterns that are *simple* and nevertheless very *powerful* and that can be applied to almost any programming language.

2.1 Closure Pattern

The most fundamental of the general programming patterns derived from ARS is the closure pattern. A closure is an **encapsulation of a lambda abstraction with its total environment**.

This environment consists of **all the names** that this lambda abstraction has access to. Access to names in a lambda abstraction is controlled by the so called *'lexical scope'*.

Lexical scope can also be described as the context of a lambda abstraction in the program text.

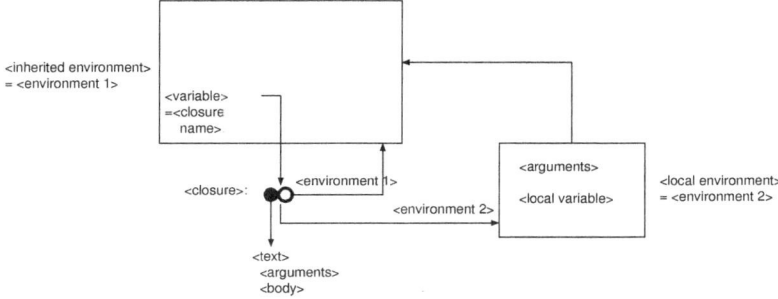

A closure is a *first class object*, which means that it can be treated like any data item:

- it can be stored in memory,
- it can be passed as an argument to a function and
- it may be returned as a value from a function.

The slight difference between a closure and an object in OOP is the following:

- A **closure** is essentially a *function* that can be called, but may have all kinds of data and procedures encapsulated in it.
- An **object** is essentially a *data item* with all kinds of data and procedures encapsulated in it. One of its procedures (normally called methods) may be 'apply', allowing to apply the object to a set of arguments, which essentially is the same as calling a function.

2.2 CLAM Pattern

The CLAM pattern is just an implementation of the general closure pattern in C.

The acronym CLAM stands for: C Lambda Abstraction. A CLAM is therefore an implementation of a lambda abstraction as defined in ARS in the programming language C.

Because a CLAM is an **encapsulation of data and procedures** (the lambda abstraction with its total environment) it can be compared to an object in object oriented technology. The easy to apply CLAM pattern derived from ARS *provides a C programmer with all the whistles and bells* functional programming and object oriented programming are justly so proud of.

2.3 List Pattern

All programming languages provide a feature to collect data in some sort of container one way or another.[1] In A++ collections are implemented as lists and dictionaries.

---
[1] In A++ no real distinction is made between data and functions, because effectively there are only abstractions:

2.3. LIST PATTERN

Lists are implemented as *linked lists of pairs*. A **pair** is composed of a *head* and a *tail*. The head points to a data item and the tail points to the next pair in the list.

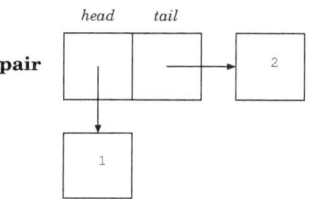

The tail in the last pair of the list points to a special element called **'nil'** or 'null'. It is the programmer's responsibility to make sure that a list is terminated with the element 'nil'.

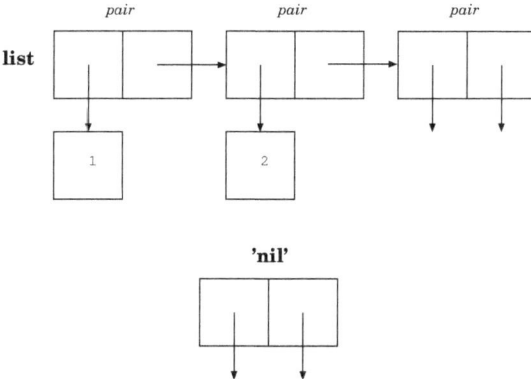

In order to work with pairs and lists the following abstractions are needed as a minimum:

- **cons** the *constructor* of a pair
- **car** the *selector* of the **head** of a pair
- **cdr** the *selector* of the **tail** of a pair

- lambda-abstractions
- references that refer to abstractions
- syntheses that return abstractions

- **nullp** a *predicate* to check, whether the list is **empty**
- **pairp** a *predicate* to check, whether the object is a pair

The section **??** on page **??** presents these abstractions and many more that can be used in the list context.

A head of a pair may be anything, i.e also a reference of another pair. This way it is possible to construct **lists of lists**, providing the power to build a **tree** or any other network structure.

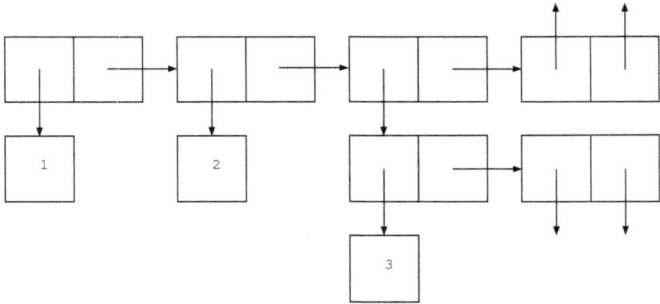

2.4 Dictionary Pattern

Dictionaries are *lists of pairs*, where the first element of the pair has the function of a **key** to the second element which is called the **value**.

Directories may be implemented as **associative lists** or as **hash tables**. Associative lists in A++ are covered in section **??** on page **??**.

2.5 Set Pattern

This pattern corresponds more or less to the list pattern with some additional functions providing set operations on lists.

2.6 Recursion Pattern

Many loop constructs in programs cover up a hidden recursive data structure. In these cases it is worth checking, whether a recursive program structure would have been the better choice.

We would like to illustrate the important programming pattern of recursion using the classical example to compute the faculty of a number, in our case the faculty of 5.

```
5! =    1 * 2 * 3 * 4 * 5
```

This expression can be rewritten like this:

```
5! =    5 * 4 * 3 * 2 * 1
```

The same holds true for the faculty of 4:

```
4! =    4 * 3 * 2 * 1
```

Comparing the expression to calculate the faculty of 5 with the expression to calculate the faculty of 4 reveals that the faculty of 4 is included in the faculty of 5. This allows us to rewrite our original expression:

```
5! = 5 * 4!
```

We may as well rewrite the expressions for the faculties of the other numbers:

```
4! = 4 * 3!
3! = 3 * 2!
2! = 2 * 1!
1! = 1
```

Using A++ notation we can write therefore:

```
(define twenty4 (mult four six))
(faculty five)
(mult five (faculty four))
(mult five (mult four (faculty three)))
(mult five (mult four (mult three (faculty two))))
(mult five (mult four (mult three (mult two (faculty one)))))
(mult five (mult four (mult three (mult two one))))
(mult five (mult four (mult three two )))
(mult five (mult four six))
(mult five twenty4)
120
```

We assume that the abstractions 'mult' for multiplication, 'add' for addition and the numbers from 1-6 have been defined. How this can be done will be shown later in section ?? on page ??.

The lines above demonstrate how the faculty of a number can be calculated by first calculating the faculty of the preceding number. This approach can of course be repeated on the next lower level until the number 1 is reached. The faculty of 1 does not have to be calculated because it is equal to 1 by definition.

2.7 Higher Order Function Pattern

This pattern is related to functions that take functions as arguments and/or return functions as values. The *classical example* for a high order function is the abstraction **'compose'**. This function takes two functions as arguments and creates a new function that is returned as value.

2.8 Message Passing Pattern

This pattern implements the *static object model* like the one normally used in Java, C++ and other OOP languages. In this model objects are created and communicated with via messages.

2.8.1 Classes of objects

A class of objects can be considered a category to which specific objects belong. Formally speaking a class defines the **attributes** an object in this category must have and the *services* that these objects are supposed to provide. Each service an object is capable of providing can be mapped to a function defined in the class, describing exactly how the service is rendered.

2.8. MESSAGE PASSING PATTERN

In OOP terminology these functions are often called **'methods'** and the attributes of an object are referred to when talking about **'instance variables'**.

The following diagram illustrates the concept of a class as introduced above[2].

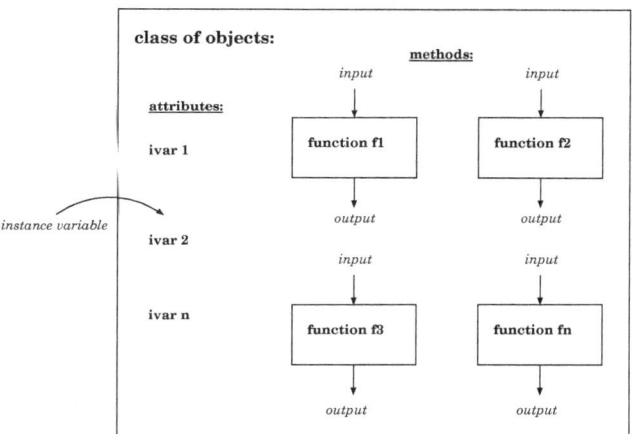

As a simple example for demonstrating the features of the 'sending messages' programming pattern we will use throughout this discussion a simplified handling of a bank account.

For this and the following examples we will illustrate the structure of a class and the relationship between the classes of a specific project using the diagramming technique recommended by Peter Coad in [6], [7], [8] and [9].

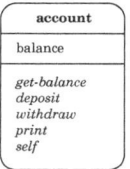

The rectangle above describing a class is split up into to three areas. The first area contains the name of the class, the second area lists the attributes of all objects of this class and the third area lists the methods corresponding to the service all objects of this class are capable to provide.

---

[2] A class in a language like C++ or Smalltalk has many other features not covered here. The features we have described are common however to all implementations of OOP technology

38   CHAPTER 2. GENERAL PROGRAMMING PATTERNS AND A++

2.8.2 Instance of a class

An instance of a class is an object having all the attributes and providing all the services defined by the class. An instance of a class is created by *invoking the constructor* of the class.

2.8.3 Constructors

In most object oriented programming languages, classes are defined by special constructs and the constructor generates an object according to the class definition. These languages can also be labelled 'class oriented'.

In **A++** however, *classes are defined entirely by constructors*, which like in other languages creates objects as instances of the specific class.

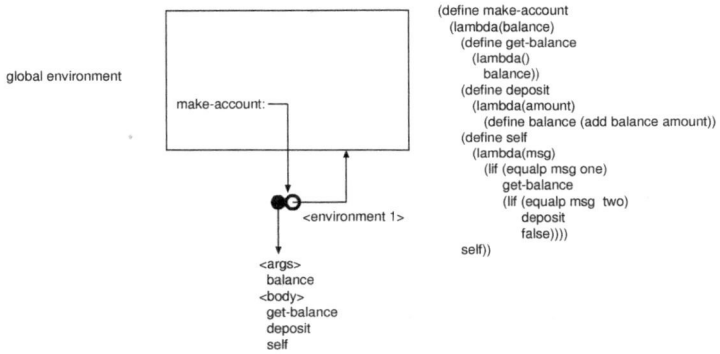

Components of a constructor

- **input arguments**
  The input arguments passed to a constructor are used to initialize the attributes. In our example the balance of the bank acount is set to 0.

- **definition of the attributes**
  Normally all attributes are defined in A++ by performing an abstraction in the first of the two possible forms, i.e. using the key word `define`.

  In our example however the input argument itself is used as an attribute. This is possible due to 'lexical scope', which, as was mentioned earlier, is coupled in A++ just like in Scheme with indefinite extent. This means it is not a variable on the runtime stack like in C and many other languages but it belongs to

2.8. MESSAGE PASSING PATTERN

the environment of the lambda-abstraction, which is preserved as long as the lambda-abstraction is still in use.

- **definition of the methods**
  The methods are of course lambda-abstractions and must be given a name using `define`.

- **definition of the special method 'self'**
  This method is a very special lambda-abstraction, because it it is returned by the constructor as the closure representing the entire object. As closure being defined and contained in the constructor it has access to everything in the constructor, i.e. all attributes and methods. This is why 'self' can serve as a dispatcher, receiving the messages sent to the object, interpreting them and mapping them to calls to the corresponding methods.

- **return value**
  The return value of the constructor is the lambda-abstraction 'self' as described above.

2.8.4 Creating instances of a class

The diagram below illustrates what happens, when a constructor is invoked to create an instance of a class. The rectangles in the diagrams represent **environment frames**. Environment frames are linked with one another making up (together with the global environent) the total environment.

The label '<environment 1>' refers to the inherited environment and '<environment 2>' to the new environment frame created by the activation of a closure (call of a function). In our example '<environment 2>' has been created by calling the constructor 'make-account'.

- The **first** environment frame created in our example represents the *top level environment* containing the constructor 'make-account' and the variable 'acc1' used later on to store the object returned by 'make-account'. An arrow pointing from the second rectangle back to this environment frame symbolizes the linkage.

- The **second** environment frame belongs to the *constructor* containing its attributes (balance) and methods (get-balance, deposit, self).

There are **two closures** of special interest in our example illustrated by the black bullets:

- the **constructor**
  taking one argument: 'balance' and having three methods. There is a line

drawn from this closure (two bullets) to the second environment labelled '<environment 2>'. This line is to be interpretet as a pointer to the closure's activation record.[3]

- the lambda-abstraction **'self'**
  representing the object returned expecting 1 argument: 'msg'. This closure has *not been invoked yet* and therefore has only the inherited environment consisting of the environment frames already discussed.

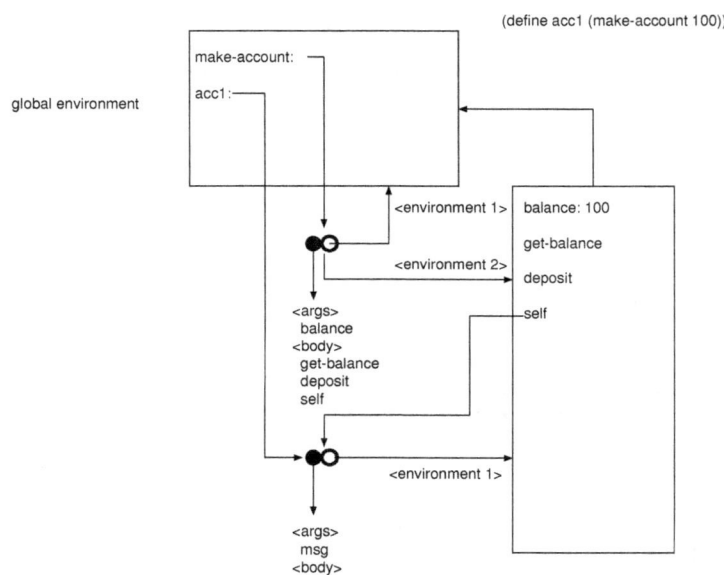

2.8.5 Sending messages

To send a message means in our context to invoke the dispatcher of the object passing the message as an argument. This was done by evaluating the expression: *(define bal-set(acc1 two))*.

Invoking the dispatcher caused the following changes in the diagram below:

- a new envionment frame was created for the dispatcher with the key/value-pair 'msg=>two' stored in it.

---

[3]The same environment is labelled '<environment 1>' viewed from the second closure however! This is no mistake. The label '<environment 1>' refers to the inherited environment and '<environment 2>' refers to the proper environment frame of a closure.

2.8. MESSAGE PASSING PATTERN

- the new environment frame was linked with the previous one
- the dispatcher interpreted the message 'two' and returned a pointer to the method 'deposit' to the caller of the function. The call was initiated from the top-level and therefore changed the top-level environment by adding the variable 'bal-set' assigning it the value returned by the call.

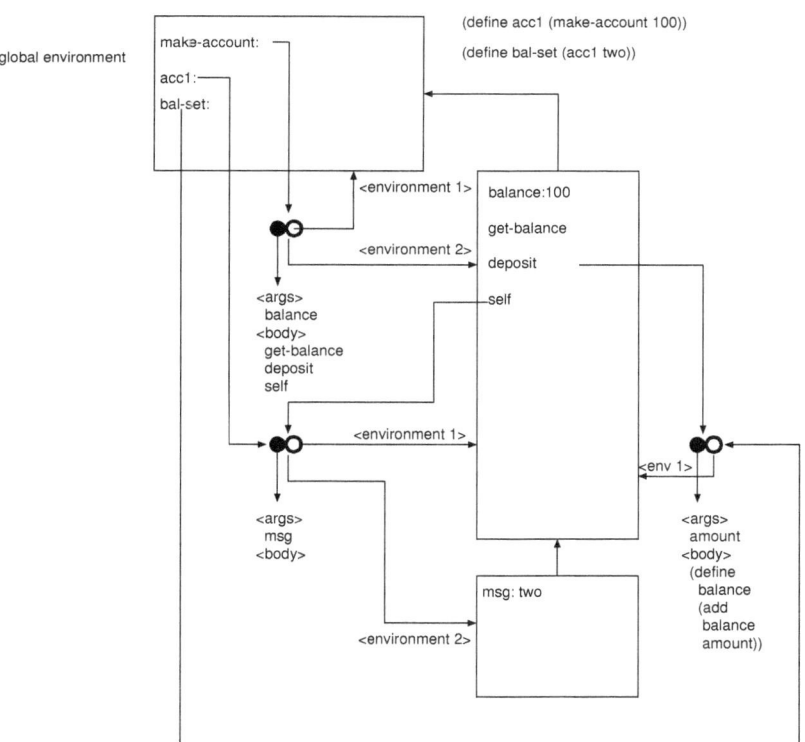

2.8.6 Executing methods

Indirectly invoking method deposit through the top-level variable 'bal-set' caused the following changes in the diagram to occur:

- the call of method 'deposit' was initiated from the top-level via the statement: *(bal-set 100)*.

42   CHAPTER 2. GENERAL PROGRAMMING PATTERNS AND A++

- a new environment frame was created for the lambda-abstraction 'deposit' with the argument of the call stored in it as a key/value=> pair: 'amount 100'.

- the new environment frame was linked with the proper environment frame of the parent of 'deposit' which is the constructor of the object.

- the arrowed line labelled '<env 1>' shows that 'deposit' has access to the environment frame of its parent.

- the body of method 'deposit' is displayed

- the return value of method 'deposit'(200) is stored in the variable 'balance' in the environment frame of the parent (constructor). The diagram shows the value of 'balance' before and after the execution of 'deposit' (100:200).

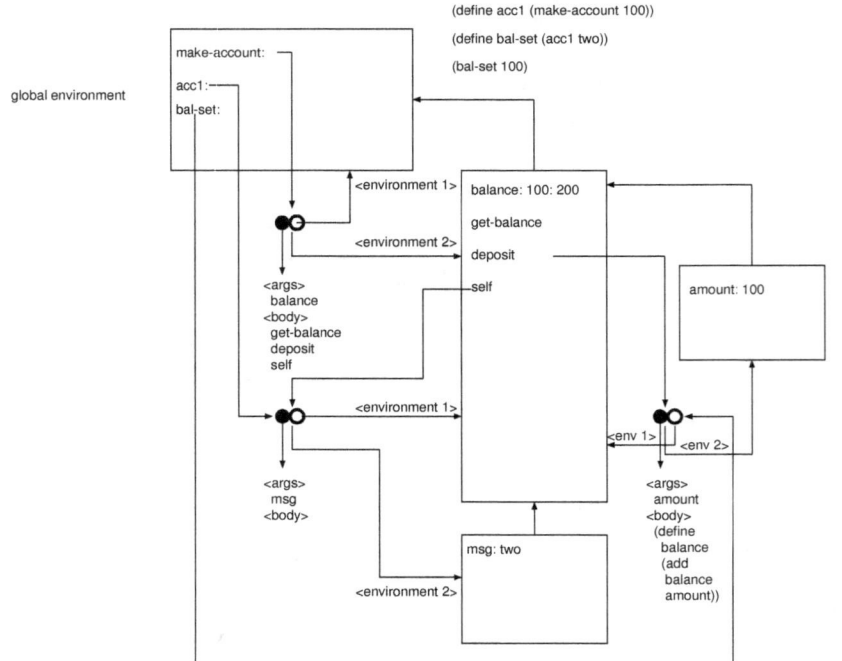

2.8.7 Essential features of object oriented programming

Encapsulation

Encapsulation in OOP-terminology means that an object has something like a shell around it protecting the attributes or status variables from being accessed from outside of the object. The belong to the object and to nobody else. Only the methods of the object have access to them, can read and modify their values. The methods, providing the services of the object, are the interface to the outside world.

In A++ this is not quite true however, because even the methods can be accessed only indirectly through the dispatcher, which was returned by the constructor when the object was created.

Inheritance

Inheritance refers to a hierarchical structure of classes as we will encounter in our second and third example of object-oriented programming in A++. In this hierarchical structure we have classes that are derived from others like the class 'dog' would be derived from 'animal' and 'car' would be derived from 'vehicle'.

Inheritance means that a derived class inherits all attributes and methods from its super class, the class from which it is derived. This feature alleviates programmers from having to start always from scratch when writing an object-oriented applications. Many times it will be possible to design and implement a class that can be derived from existing ones in some sort of class library, which one has either written oneself or that can be obtained from another source.

In A++ in its original form inheritance is not offered as a language feature identical to the one described above, because there is no predefined syntax to define classes. It would most likely be possible to define abstractions implementing such a class system. Nobody has done it yet, because there is a very simple mechanism available that can be considered functionally equivalent to inheritance. It is called **'delegation'**. In our second example of object-oriented programming we will implement the inheritance using delegation.

Delegation can be very briefly explained like this: An object of a derived class receives a message from a client. If the object is equipped to provide the service it will do it. If the object does not have the proper methods to respond to the message it will delegate the message to another object, which is an instance of its super class. Messages are this way delegated up in the hierarchy until nobody can answer it, which would result in an error return.

Peter Coad compares delegation with inheritance in [6] and draws the conclusion that delegation should be favorized (chapter 2, page 49).

Polymorphism

Polymorphism in OOP-Terminology means that a certain message may be implemented by different objects in different ways. A good example is the 'display'-message sent to a number, a string of characters or a list. It is always the message 'display yourself!'. The object must know itself what has to be done and how, nobody cares as long as the object appears on the screen.

2.8.8 Relation between classes

The duality of relations below corresponds to the duality 'inheritance' and 'delegation' we have just discussed.

- **IS-Relation**
  This relation has to do with inheritance: A dog *is* an animal. A car *is* a vehicle. Because a dog is animal it has all the features of animal and of course on top of that its own.

- **HAS-Relation**
  In the system using delegation described above we have seen that an object must have an instance of its super-class in its attributes in order to be able to delegate an unknown message to a higher level.

  Of course an object may have instances of several super-classes making it easy to implement multiple inheritance.

2.9 Meta Object Protocol Pattern

The pattern implements a *dynamic object model*. In a this model methods can be dynamically created, modified and deleted. Such a dynamic object model is available in Common Lisp under the name of CLOS. In Scheme it is very easy to create your own static as well as dynamic object system. In A++ it was most likely never attempted but would not be impossible.

Bibliography

[1] Harold Abelson and Gerald Jay Sussman with Julie Sussman. *Structure and Interpretation of Computer Programs*. The MIT Press, Cambridge, Massachusetts, second edition, 1996.

[2] H. Barendregt. *The Lambda Calculus – Its Syntax and Semantics*. North-Holland, Amsterdam, 1981.

[3] Jacques Chazarain. *Programmer avec Scheme – De la pratique à la théorie*. International Thomson Publishing France, Paris, 1996. ISBN 2 84180 130 4.

[4] Tom Chritiansen and Nathan Torkington. *Perl Cookbook*. O'Reilly, Sebastopol, CA, 2003.

[5] Alonzo Church. *The Calculi of Lambda Conversion*. Princeton University Press, Princeton, New Jersey, 1941.

[6] Peter Coad and Mark Mayfield. *Java Design – Building Better Apps & Applets*. Yourdon Press, Upper Saddle River, New Jersey, 1997.

[7] Peter Coad and Jill Nicola. *Object-Oriented Programming*. P T R Prentice-Hall, Inc., Englewood Cliffs, New Jersey, 1993.

[8] Peter Coad and Edward Yourdon. *Object-Oriented Analysis – Second Edition*. Prentice-Hall, Inc., Englewood Cliffs, New Jersey, 1991.

[9] Peter Coad and Edward Yourdon. *Object-Oriented Design*. Prentice-Hall, Inc., Englewood Cliffs, New Jersey, 1991.

[10] Douglas Crockford. *Das Beste an JavaScript*. O'Reilly Verlag GmbH&Co.KG Inc., Balthasarstr. 81, 50670b K"oln, 2008.

[11] Douglas Crockford. *JavaScript – The Good Parts*. O'Reilly Media Inc., 1005 Gravenstein Highway North, Sebastopol, CA 95472, 2008.

[12] Howard B. Curry and R. Feys. *Combinatory Logic*, volume 1. North-Holland, Amsterdam, 1958.

[13] R. Kent Dybvig. *The Scheme Programming Language*. Prentice Hall, Englewood Cliffs, New Jersey, 1987.

[14] R. Kent Dybvig. *The Scheme Programming Language*. Prentice Hall, Upper Saddle River, New Jersey, zweite edition, 1996.

[15] Matthias Felleisen and Daniel P. Friedman. *A Little Java, A Few Patterns*. The MIT Press, Cambridge, Massachusetts, 1998.

[16] Iain Ferguson. *The Schemer's Guide*. Schemers Inc., Fort Lauderdale, Florida, 1995.

[17] Daniel P. Friedman and Matthias Felleisen. *The Little Schemer – Fourth Edition*. The MIT Press, Cambridge, Massachusetts, 1996.

[18] Daniel P. Friedman and Matthias Felleisen. *The Seasoned Schemer*. The MIT Press, Cambridge, Massachusetts, 1996.

[19] Brian Harvey and Matthew Wright. *Simply Scheme – Introducing Computer Science*. The MIT Press, Cambridge, Massachusetts, 1994. ISBN 0 262 08226 8.

[20] Peter Henderson. *Object-Oriented Specification And Design with C++*. McGraw-Hill Book Company, London, 1993.

[21] Simon L. Peyton Jones. *The Implementation of Functional Programming Languages*. Prentice Hall International, Hertfordshire,HP2 7EZ, 1987. ISBN 0 13 453325 9.

[22] Samuel N. Kamin. *Programming Languages – An Interpreter-Based Approach*. Addison-Wesley Publishing Company, Reading, Massachusetts, 1990. ISBN 0 201 06824 9.

[23] Richard Kelsey, William Clinger, and Jonathan Rees (editors). Revised[5] report on the algorithmic language scheme. *Higher-Order and Symbolic Computation*, 11(1):7–105, 1998.

[24] Stephen G. Kochan. *Programming in C*. Hayden Books, Indianapolis, Indiana, 1988.

[25] Stephen G. Kochan and Patrick H. Wood. *Topics in C Programming*. Hayden Books, Indianapolis, Indiana, 1988.

[26] Martin Kstner. *Perl frs Web*. Galileo-Press GmbH, Bonn, Germany, 2003.

[27] Georg P. Loczewski. *Programmierung PUR – Programmieren fundamental und ohne Grenzen*. S.Toeche-Mittler Verlag, 64295 Darmstadt, 2003. ISBN 3 87820 108 7.

[28] Georg P. Loczewski. *A++ Die kleinste Programmiersprache der Welt - Eine Programmiersprache zum Erlernen der Programmierung*. tredition Verlag, Hamburg, 2018. ISBN 978-3-7469-3099-2.

[29] Georg P. Loczewski. *A++ The Smallest Programming Language in the World*. tredition Verlag, Hamburg, 2018. ISBN 978-3-7469-3021-3.

[30] Peter Norvig. *Paradigms of Artificial Intelligence Programming: Case Studies in Common Lisp*. Morgan Kaufmann Publishers, San Francisco, California, 1992.

[31] Jean-Christophe Routier and Eric Wegrzynowski. *Débuter la programmation avec Scheme*. International Thomspn Publishing France, Paris, 1997.

[32] Michael Schilli. *Effektives Programmieren mit Perl 5*. Addison Wesley Longman GmbH, Bonn, Germany, 1997.

[33] George Springer and Daniel P. Friedman. *Scheme and the Art of Programming*. The MIT Press, Cambridge, Massachusetts, 1993. ISBN 0 262 19288 8.

[34] J.E. Stoy. *Denotational Semantics*. MIT Press, Cambridge, Massachusetts, 1981.

[35] Glenn L. Vanderburg and andere Autoren. *Tricks of the Java Programming Gurus*. Sams.net Publishing, Indianapolis, Indiana, 1996.

[36] Larry Wall, Tom Christiansen, and Randal L. Schwartz. *Programming Perl*. O'Reilly & Associates, Inc., Sebastopol, CA, 1996.

The World of ARS

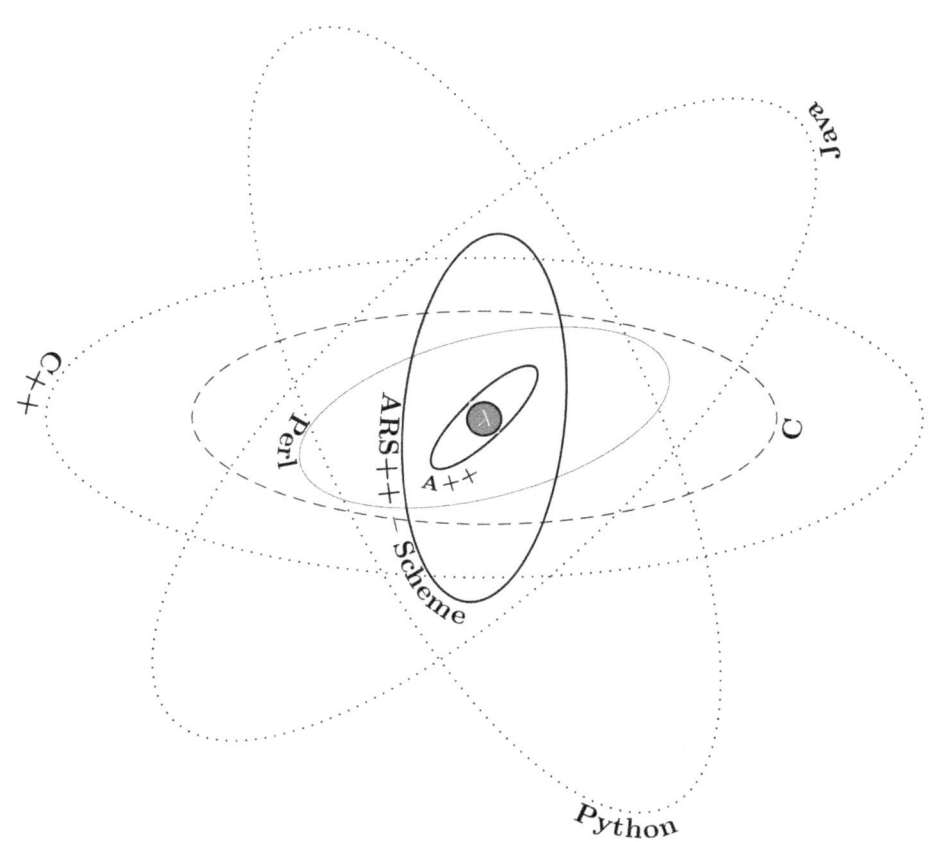

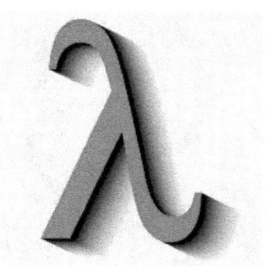

A++
The Smallest Programming Language in the World
An Educational Programming Language
ISBN 978-3-7469-3811-0 (Paperback)
http://www.alpha-bound.de
http://www.tredition.de

www.ingramcontent.com/pod-product-compliance
Lightning Source LLC
Chambersburg PA
CBHW071217240526
45470CB00018B/2071